Lean Analytics

The Complete Guide to Using Data to Track, Optimize and Build a Better and Faster Startup Business

Jeffrey Ries

© **Copyright 2018 by Jeffrey Ries. All rights reserved.**

This document is geared towards providing exact and reliable information regarding topic and issue covered. The publication is sold with the idea that the publisher is not required to render accounting, officially permitted, or otherwise, qualified services. If advice is necessary, legal or professional, a practiced individual in the profession should be ordered.

From a Declaration of Principles which was accepted and approved equally by a Committee of the American Bar Association and a Committee of Publishers and Associations.
In no way is it legal to reproduce, duplicate, or transmit any part of this document in either electronic means or in printed format. Recording of this publication is strictly prohibited and any storage of this document is not allowed unless with written permission from the publisher. All rights reserved.

The information provided herein is stated to be truthful and consistent, in that any liability, in terms of inattention or otherwise, by any usage or abuse of any policies, processes, or directions contained within is the solitary and utter responsibility of the recipient reader. Under no circumstances will any legal responsibility or blame be held against the publisher for any reparation, damages, or monetary loss due to the information herein, either directly or indirectly.

Respective authors own all copyrights not held by the publisher.
The information herein is offered for informational purposes solely, and is universal as so. The presentation of the information is without contract or any type of guarantee assurance.

The trademarks that are used are without any consent, and the publication of the trademark is without permission or backing by the trademark owner. All trademarks and brands within this book are for clarifying purposes only and are the owned by the owners themselves, not affiliated with this document.

Table of Contents

Introduction 7

Chapter 1: What is Lean Analytics? 8
    What is Lean? 9
    Lean Analytics 10
    Focus on the fundamentals 11
    Waste and the Lean System 12
    How Lean can help you define and then improve a value stream 14

Chapter 2: The Lean Analytic Stages Each Company Needs to Follow 16

Chapter 3: The Lean Analytics Cycle 19
    What do I need to improve? 20
    Form a hypothesis 21
    Conduct an experiment 22
    Measure your outcomes and make a decision 23

Chapter 4: False Metrics vs. Meaningful Metrics 25
    The biggest false metrics to watch out for 26

Chapter 5: Recognizing and Choosing a Good Metric 31
Types of Metrics 32

Chapter 6: Simple Analytical Tests to Use 35
Segmentation 35
Cohort Analysis 39
A/B Tests 40

Chapter 7: Step 1 of the Lean Analytical Process: Understanding Your Project Type 43
E-commerce 44
The best metrics 45
Software as a service 47
The best metrics 48
Mobile app companies 49
Best metrics 50
Media site companies 51
Best metrics 52
User-generated content business 53
The best metrics 54
Two-sided marketplace business 55
Best metrics 56

Chapter 8: Step 2: Determine Your Current State...58
   Stage 1: Empathy or is this a real problem? 58
   Stage 2: Stickiness or do I have a good solution? 59
   Stage 3: Virality or does this solution provide value to enough people? 60
   Stage 4: Revenue or can I make this profitable? 61
   Stage 5: Scale or can we expand to a bigger audience? 61
   What can I do with these innovation stages? 62

Chapter 9: Step 3: Pinpoint the Most Pressing Metric 64
   How can I find that one metric? 65
   What to do after optimizing that one metric 66

Chapter 10: Tips to Make Lean Analytics More Successful for You...68
Conclusion 72

Introduction

Congratulations on getting a copy of *Lean Analytics: The Complete Guide to Using Data to Track, Optimize and Build a Better and Faster Startup Business* and thank you for doing so.

The following chapters will cover everything you need to know to get started with the process of Lean Analytics. The Lean Support system is a great way to ensure that your business is as efficient as possible by eliminating the amount of waste that is present. The Lean Analytics section is going to help with data collection and analysis. Thus, you'll determine where the waste is present, and this will help you to pick the right metrics to implement.

This book will discuss Lean Analytics and how its processes can help you reduce waste and find the best strategy to improve your business. It is just one step in the Lean Support System, but it is an extremely critical step. This guidebook will provide you with the information that you need to get started so that you can become an expert in Lean Analytics in no time.

There are plenty of books on this subject on the market, so thanks again for choosing this one! Please enjoy!

Chapter 1: What is Lean Analytics?

The central idea behind Lean Analytics is on enabling a business to track and then optimize the metric that will matter the most to their initiative, project, or current product.

There is often a myriad of methods to improve your product, but you may not have the time to work on all of them. With Lean Analytics, you will learn how to find and address the one thing that will make the biggest difference.

Setting the goal of focusing on the right method will help you see real results. Just because your business has the ability and the tools to track many things at once, does not mean that it would be in your best interest to do so.

Tracking several types of data simultaneously can be a great waste of energy and resources and may distract you from the actual problems. Instead, you will want to focus your energy on determining that one vital metric. This metric will make the difference in the product or service that you provide.

The method in your search for this metric will vary depending on your field of business and several other factors. The way that you'll find this metric is through an in-depth understanding of two factors:

- The business or the project on which you're presently working.
- The stage of innovation that you are currently in.

Now that we have a basic understanding of Lean Analytics and what it means let's take some time to further explore and see its different parts.

What is Lean?

Lean is a method that is used to help improve a process or a product on a continuous basis. This works to eliminate the waste of energy and resources in all your endeavors. It is based on the idea of constant respect for people and your customers, as well as the goal of continuously working on incremental improvements to better your business.

Lean is a methodology that is vast and covers many aspects of business. This guidebook will spend sufficient time discussing a specific part of Lean, Lean Analytics. Here, you can learn how to make the right changes. Of course, you will need a working understanding of where to start, and Lean Analytics can help.

Lean is a method that was originally implemented for manufacturing. The idea was to try to eliminate wastes of all kinds in a business, allowing them to provide great customer service and a great product while increasing profits at the same time. Despite its beginnings, the Lean methodology has expanded to work in almost any kind of business. As long as you provide a product or a service to a customer, you can use the Lean methodology to help improve efficiency and profits.

For instance, how will you determine which metric will help you succeed? Which metric will prove to be the best and result in the most improvement compared to others? How will the metric help, how should it be implemented, and how can you ascertain if it's successful in the end? Lean Analytics can help you gather the necessary information to find and work with the right metric.

Lean Analytics

Lean Analytics is part of the methodology for a lean startup, and it consists of three elements: building, measuring, and learning. These elements are going to form up a Lean Analytics Cycle of product development, which will quickly build up to an MVP, or Minimum Viable Product. When done properly, it can help you to make smart decisions provided you use the measurements that are accurate with Lean Analytics.

Remember, Lean Analytics is just a part of the Lean startup methodology. Thus, it will only cover a part of the entire Lean methodology. Specifically, Lean Analytics will focus on the part of the cycle that discusses measurements and learning.

It is never a good idea to just jump in and hope that things turn out well for you. The Lean methodology is all about experimenting and finding out exactly what your customers want. This helps you to feel confident that you are providing your customers with a product you know they want. Lean Analytics is an important step to ensuring that you get all the information you need to make these important decisions.

Before your company decides to apply this methodology, you must clearly know what you need to track, why you are tracking it, and the techniques you are using to track it.

Focus on the fundamentals

There are several principles of Lean that you will need to focus on when you work with Lean Analytics. These include:

- A strive for perfection
- A system for pull through
- Maintain the flow of the business
- Work to improve the value stream by purging all types of waste

- Respect and engage the people or the customers
- Focus on delivering as much value to the customer as effectively as possible

Waste and the Lean System

One of the most significant things that you will be addressing with Lean Analytics, or with any of the other parts of the Lean methodology, is waste. Waste is going to cost a company time and money and often frustrates the customer in the process. Whether it is because of product construction, defects, overproduction, or poor customer service, it ends up harming the company's bottom line.

There are several different types of waste that you will address when working with the Lean system. The most common types that you will encounter with your Lean Analytics include:

- **Logistics:** Take a look at the way the business handles the transportation of the service or product. You can see if there are is unnecessary movement of information, materials, or parts in the different sections of the process. These unnecessary steps and movements can end up costing your business a lot of money, especially if they are repeated on a regular basis. This will help you see if more efficient methods exist.

- **Waiting:** Are facilities, systems, parts, or people idle? Do people spend much of their time without tasks despite the availability of work or do facilities stay empty? Inefficient conditions can cost the business a lot of money while each part waits for the work cycle to finish. You want to make sure that your workers are taking the optimal steps to get the work done, without having to waste time and energy.

- **Overproduction:** Here, you'll need to take a look at customer demand and determine whether production matches this demand or is in excess. Check if the creation of the product is faster or in a larger quantity than the customer's demand. Any time that you make more products than the customer needs, you are going to run into trouble with spending too much on those products. As a business, you need to learn what your customer wants and needs, so you make just the amount that you can sell.

- **Defects:** Determine the parts of the process that may result in an unacceptable product or service for the customer. If defects do exist, decide whether you should refocus to ensure that money is not lost.

- **Inventory:** Take a look at the entire inventory, including both finished and unfinished products. Check for any pending work, raw materials, or finished goods that are not being used and do not have value to them.

- **Movement:** You can also look to see if there is any wasted movement, particularly with goods, equipment, people, and materials. If there is, can you find ways to reduce this waste to help save money?
- **Extra processing:** Look into any existing extra work, and how much is performed beyond the standard that is required by the customer. Extra processing can ensure that you are not putting in any more time and money than what is needed.

How Lean can help you define and then improve a value stream

Any time that you look at the value stream, you will see all the information, people, materials, and activities that need to flow and cooperate to provide value to your customers. You need these to come together well so that the customer gets the value they expect, and at the time and way, they want it. Identifying the value stream will be possible by using a value stream map.

You can improve your value stream with the Plan-Do-Check-Act process. This strategy can be used upfront so that you can design the right processes and products before they reach their finished form. Additionally, the strategy helps you to create an environment that is safe and orderly and allows easy detection of any waste.

==Another method of creating this environment is the 5S+ (Five S plus): sort, straighten, scrub, systematize, and standardize.== Afterward, ensure that any unsafe conditions along the way are eliminated.

The reason that you will want to do the sorting and cleaning is to make it easier to detect any waste. When everything is a mess, and everyone is having trouble figuring out what goes where, sorting and cleaning can address waste quite fast. There will also be times when you deem something as waste and then find out that it is actually important.

When everything is straightened out, you can make more sense of the processes in front of you. Afterward, you can take some time to look deeper into the system and eliminate anything that might be considered as waste or unsafe, and spend your time and money on parts of the process that actually provide value for your customer.

Chapter 2: The Lean Analytic Stages Each Company Needs to Follow

To be successful with Lean Analytics, you'll need to follow several different stages. You won't be able to move on to the next stage if you do not complete the preceding step. There are five in particular that you will need to focus on to get work done with this section of the Lean support methodology. The five stages are:

- **Stage 1:** The initial stage is where you will concentrate on finding the problem for which people are searching for a solution. A business that focuses on business to business selling is going to find this stage critical. When you address this problem, then you can move on to the next stage.
- **Stage 2:** For this stage, you are going to create an MVP product that can be used by early adopter customers. This stage is where you are aiming for user retention and engagement, and you can spend some time learning how this will happen when people start to use the product. You can also learn this information based on how the customer uses your site and how long they stay. You'll take some time at this stage because you will need

to experiment and also may need to go through and choose from a few different products before you get the one that is right for you. Once you have this information, you can move on.

- **Stage 3:** Once you find out how the early adopter customers are going to respond to a product or service, it is time to find the most cost-efficient way to reach more customers. Once you have a plan ready to get those customers, and then more of them start purchasing the product, then you can move to the next stage. You would not want to go with a product that may be popular but costs a ton of money, which will cut into your revenues and can make it difficult to keep growing in the future.

- **Stage 4:** You are now going to spend some time on economics and focusing on how much revenue you are making. You want to be able to optimize the revenue, so you need to calculate out the LTV:CAC ratio. LTV is the revenue that you expect to get from the customer, and the CAC is the cost that you incurred to acquire that customer. You can find this ratio by dividing your LTV by the CAC. Your margins are doing well if you get an LTV that is three times higher than the CAC. The higher the margins you get, the better because that means you are going to earn more in profits from the endeavor.

- **Stage 5:** In the final stage, you will then take actions that are necessary to grow the business. You can continue with your current plan if you are making a high enough margin from the previous steps, or you may need to make some changes to ensure that you can earn enough revenue to keep the business growing. You can also spend time making plans on where you would like to concentrate on in the future to increase the growth of your business and help it expand. The main goal for your business is to keep growing and increase revenue. This step helps you to reevaluate what you have in your current plan and decide if it is working for you or if you need to go with a different option.

Chapter 3: The Lean Analytics Cycle

The Lean Analytics Cycle is vital in helping you get started on this part of the Lean support methodology with your business. There are four steps that will come with this process, and following each one can be crucial in ensuring that this works for you.

The best way to think about the Lean Analytics Cycle is like the scientific method. You need to do some thinking to determine what needs to be improved in your business, form a hypothesis to help lead your findings, and then perform experiments to see if that is the right process for you to keep following. If things don't work out, you don't just give up. You will continue to find new experiments, going with the same hypothesis if it works (otherwise you'll need to form a new hypothesis) until you find the right solution.

The Lean Analytics Cycle will be incredibly helpful when you begin going through the entire process. Let's take a look at the steps that you need to fulfill to use the Lean Analytics Cycle.

What do I need to improve?

Before you can do anything with the Lean Analytics Cycle, you must really understand your business. You need to know all the important aspects of your business, in addition to knowing what you want to change.

During this first step, you may need to talk to other businessmen to help you find what metric you should use, based on what is most relevant to your business right now. You may also want to take a look at your business model to find out what metric will work best for you.

After you have time to choose a metric, you should connect it to the KPI or the Key Performance Indicator. An example of this is the metric that is seen as a conversion rate if the KPI is the number of people who currently purchase the product.

To make this step easier, the first thing that you would want to do is write down three metrics that are important for your business. Afterward, write down the KPI that would be measured for each metric.

Never try to implement the Lean system without understanding the most important processes that need to be improved. Sure, you could probably make a long list of things that you may want to improve in your business. But you won't really see the benefits of the Lean system if you don't pick things that are important to the overall functioning of your business. Look closely at what your business needs to improve, and pick the one that is the most important before moving on.

Form a hypothesis

This is a stage where a level of creativity needs to come into play. The hypothesis is going to give you the answers that you need to move forward. You will need to look for inspiration, and you can find it in one of two ways. You can look for an answer for something like "If I perform ___, I believe ____ will happen, and ____ will be the outcome."

The first place you can look into is any data that you have available. Often, this data will provide you with the answer that you need. If you do not have data at all, you may need to do some studying of your own to come up with an answer. You could use some of the strategies from your competitors, follow the practices that have worked well for others, do a survey, or study the market to see what the best option will be.

What you need to keep in mind here is that the hypothesis is there to help you to think like your audience. You want to keep asking questions until you understand what they are thinking, or learn to understand the behavior of your audience or customer.

Conduct an experiment

After you have taken the time to form a hypothesis, it is time to test it out with the help of an experiment. There are three questions that you need to consider to get started with an experiment:

- **Who is the target audience?** You need to carefully consider who your customers are and whether or not they are the right customers, or if you should look somewhere else to get better results. Also, think about some of the ways that you reach them, and if there are better ways to do this.
- **What do you expect the target audience to do?** This often includes purchasing the product, using the product, or something similar. You can then figure out if the audience understands what you want them to do; is it easy for them to do this action, and how many of the target audience are completing the task?
- **Why do you think they should accomplish the action**? Are you providing them with the right

motivation to accomplish the task? Do you think that the strategy is working? If they aren't being motivated enough by you, are they doing these things for the competitors or otherwise?

Answering these questions is vital because they may help you understand your customer better than ever before. Creating your experiment during this stage does not have to be difficult. Try using the following sentence to help you get started:

"WHO will do WHAT because WHY to improve your KPI towards the defined goals or target."

If you have gone through and come up with a good hypothesis in the previous step, then it shouldn't be too hard to create a good experiment as well. Then, once you have the experiment, you can go through and set up the Lean Analytics so that you can measure your KPI and carry on in the experiment.

Measure your outcomes and make a decision

You can't just get started with an experiment and then walk away from it. You need to measure how well it goes to determine if it is truly working; if some changes are needed; or if you need to work from scratch. You can then make a decision on the next steps you need to take. Some of the things to look for when measuring the outcomes during this stage include:

- **Was the experiment a success?** If it is, then the metric is done. You can move on to finding the next metric to help your business.
- **Did the experiment fail?** Then it is time to revise the hypothesis. You should stop and take some time to figure out why the experiment failed so that you have a better chance at a good hypothesis the next time.
- **The experiment moved but was not close to the defined goal.** In this scenario, you will still need to define brand new experiment. You can stay with the hypothesis if it still seems viable, but you would need to change up the experiment.

Chapter 4: False Metrics vs. Meaningful Metrics

One of the prerequisites for working with Lean Analytics is understanding that most people are using their data wrong. When you don't use your data correctly, you are not going to be able to come up with patterns, opportunities, or results that are achievable.

There are two points that come with this idea. These two points are:

- There are many companies, as well as people, who will label themselves with descriptions like "data-driven." Sure, they may use up a lot of their resources on compiling data. However, they then miss out on the "driven" part. Few are actually going to base a strategy on the information that they acquire from the data. They may have the right data, but they either don't understand it or choose to react to it incorrectly.
- Even if the actions of a company or person are driven by data, the problem of using wrong data still exists. Often, they will oversimplify these metrics and then use them according to the convention. Keep this in mind: just

because other people do this or have done this doesn't mean it is going to prove useful to the goals that you have. Consequently, the data is going to become garbage in, and then the analysis is garbage out. This is often known as false data.

As a business who is interested in working with Lean Analytics, it is important to learn the difference between false metrics and meaningful metrics. If you follow false metrics, you are going to be following a strategy that is not going to help you reach your goals, which will mean a lot of time, effort, and resources wasted.

The biggest false metrics to watch out for

As a business that is trying to cut out waste and ensure that you provide the best customer service and the best products possible, you must always ensure that you watch out for some of the false metrics that may come up. Many people who don't understand how data works will be taken in by these false metrics that, in reality, will mean wasted time and resources. Some of the most common false metrics for you to watch out for include:

- **The number of hits:** Just because you have a website that is attractive and contains many points of interest, doesn't necessarily mean that it will tell you what the users are really interested in. You should not focus on the number of hits your website gets. This may make you feel good about your website, and it can be neat to see how many people come and visit your website. But you need to focus more on what the customer is interested in or looking for.
- **Page views:** This metric refers to how many pages are clicked on a site during a given time. This is a slightly better than hits, but you typically don't want to waste your time with this metric. In most cases, unless you are working with a business that does depend on page views, such as advertising, the better metric for you to use is to count people. You can do this with tools that will provide information on unique visitors per month.
- **Number of visitors:** The biggest problem with this metric is that it is often too broad. Does this type of metric talk about one person who visited the same site a hundred times, or a hundred people who visited once? You most likely want to look at the second group of people because you've obtained more impressions. Otherwise, just looking at the number of visitors will not give you this information.

- **Number of unique visitors:** This is a metric that is going to tell you how many people got to your website and saw the home page. This may sound good at first, but it is not going to give you any valuable information. You may also want to find out things like how many visitors left right away when they saw the page or how many stayed and looked around. Unique visitors can help you see that some new people are coming onto your website and checking things out but they don't really tell you much about those visitors or what they are doing.
- **Number of likes, followers, or friends**: This is a good example of a vanity metric that is going to show you some false popularity. A better metric that you can go with is the level of influence that you have. What this means is how many people who will do what you want them to do. While it is good to have followers and likes on your page to show that people are looking at your content, it is not as important as some of the other metrics that you can pick.
- **Email addresses:** Having a big list of email addresses is not a bad thing by itself. But just because you have this large list does not mean that everyone on it is going to open, read, and act on the messages that you send out. You want to make sure that the email addresses that you do have are high quality and are from people who actually want to hear from you, even if that means

your email list is a little bit smaller. If you are collecting emails, strive to get addresses from people who are actually interested in your product and services. Don't just collect emails so you can boast of a large list.

- **The number of downloads:** This is a common metric that is used for downloadable products. While it can help with your rankings in the marketplace when you are in the app store, the download number is not going to tell you anything in depth, and it won't give you any real value. If you would like to get some precise answers here, you can pick some better metrics. The Launch Rate is a good place to start because it will show the percentage of those who downloaded, created an account, and then used the product. You can also use something like Percentage of Users Who Pay so you can see how many actually pay for anything.

- **Time spent by customers on a page or website:** The only time that this is going to be useful is for businesses that are tied directly to the behavior of the engaged time. For example, a customer could spend a lot of time on your web page, but what if they are spending that time on the help pages or on the complaints pages? This is not necessarily a good thing for your business, so this metric is not the best one for you to go with.

If you want to pick out a metric that will actually help your business get ahead, then you must make sure that you avoid some of these false metrics. They may look good on the surface, but in reality, they are just giving you information that could be pretty useless, and you will end up wasting a lot of time and money to follow them.

Chapter 5: Recognizing and Choosing a Good Metric

Part of the Lean Analytics methodology is finding a good metric to help you out. The Lean Analytics Cycle is a measurement of movement towards a goal that you already defined. So, once you have taken the time to define your business goals, then you must also think about the measurements you can make to progress towards the goals.

This can be hard to do. How are you supposed to find a good metric that can make sure you go towards the goals that you set out? Some of the characteristics that you can look for when searching for a good metric include:

- **Comparable:** You know that you have a metric that is good if it is comparable. You want to be able to compare how things have changed in the last year, or even from one month to another. This gives you a good idea if there have been any changes, positive or negative, with your business process, customer satisfaction, and more. You can ask yourself these questions about the metric to help test for this:

- How was the metric last year, or even last month?
- Is the rate of conversion increasing? You can use the Cohort Analysis to help with conversion rate tracking.

- **Understandable:** The metric that you use should never be complex or complicated. Everyone should be able to understand what it is. This ensures that they know what the metric is measuring.
- **Ratio:** You should never work with absolute numbers when you are working with metrics. If you find that you have those, you should try to convert them to make comparisons easier, which in turn makes it easier to make decisions.
- **Adaptability:** If you have chosen a good metric, it should change the way that the business changes. If you notice that the metric is moving, but you have no idea why it is moving, then it is never a good metric. The metric should move with you, not randomly on its own, or it won't be a secure one to use.

Types of Metrics

There are two metric types that you are able to use when doing Lean Analytics. These include qualitative and quantitative metrics. To start, qualitative means that the metric has a direct contact with your customers. This would be things such as feedback and interviews. It is going to provide you with some detailed knowledge of the metric.

You can also work with quantitative metrics. These are more of a number form of metrics. You can use these to ask the right types of questions from the customer.

Of course, both of these methods have other things under them that make them easier to use. You will find that both of these methods have actionable and vanity metrics.

- **Vanity metrics** will not end up changing the behavior of the thing you are concerned about. These are a big waste of your time, and you should avoid them as much as possible. They seem to provide you with some good advice and something that you can act upon, but often they don't lead you anywhere and can make things more difficult. If you are working with a company to help you determine your metrics, be very wary if they start touting the benefits of following any of the vanity metrics.
- **Actionable metrics** are going to end up changing the behavior of the thing you are concerned about. These are the types of metrics that you want to work with on your project. They are metrics that can lead you to the plan that you should follow and can make it easier to come up with a strategy to make your business more efficient.

- **Reporting metrics** is a good way to find out how well the business is performing when it does even everyday activities.
- **Exploratory metrics** are going to be useful for helping you to find out any facts that you do not know about the business.
- **Lagging metrics** are good to work with when you want more of a history of the organization and you want as many details as possible to help with a decision. The churn of a company can be a good example of the lagging metrics. This is because it is going to show you how many customers have canceled their orders for a specific amount of time.
- **Leading metrics** are good because they can help provide you with the information that you need to make future forecasts for the business. Customer complaints can be a good example of leading metrics because it can help you to predict how a customer will react.

You will need to determine which kind of metric you want to use based on the problem or project that you are working on. Working with one metric is usually best. Doing so will help keep you on track, so you know what to look for. Don't waste your time trying to work on more than one metric. You will only get confused and end up with no clear idea about the strategy to follow.

Chapter 6: Simple Analytical Tests to Use

Another thing that you should concentrate on to do well with Lean Analytics is to have some familiarity with the tests that are used. These tests are helpful because they are going to be used to help you examine any assumptions that you are trying to use here. These tools can also be used to help you identify customer feedback so you can respond to them properly. Let us take a look at some of the best analytical tests that you can use when working with Lean Analytics.

Segmentation

The first test is segmentation. This process involves comparing a set of data from a demographic bucket. You can divide up the demographics in any manner that you choose such as gender, lifestyle, age, or where they live. You can use this information to find out where people are purchasing a particular product; if there are different buying behaviors between female and male customers; and if your target audience seems to be in a certain age group or not.

The reason that you want to build up a user segment is to make it easier for the data to be actionable. Analytics can teach you a ton about the people who purchase from you, but there is often a lot of information there, and it can be hard to draw good conclusions from this information. After all, while this information from the past can be useful, it isn't going to be the best to tell you how to improve either retention or conversion rate.

This is where the process of segmentation is going to come into play. When you learn how to ==filter out the audience, you will then be better able to create a plan to make new products that serve them the most==. Analytics can give you the information that you need, but segmentation can help you to act.

For example, you may have a conversion rate that seems average or good, but it could be from a combination of one group that converts really high and consistently so, and then another group that seems to never convert at all. You could be wasting a lot of money on that second group where you are hardly getting anybody to convert at all. Segmentation can be used to help you understand what things you are doing the right way when engaging the first group, and can give you a plan on how you can improve to work on that second group.

With segmentation, you don't want to only look at the data to learn some more about your users, but you also want to come up with data that you can act upon. Segmentation can help you with this. You will be able to ==divide up the people in your customer base and learn how to advertise to them better than ever before.==

Remember that not all customers are going to be the same. There are some of your customers who may purchase something once, and they aren't regular customers. While it is still good to reach out to them, you want to learn who your regular audience is, what they respond to, and what keeps them coming back. This is going to ensure that you keep them coming back and earn as much profit as possible.

So, how do you create a segment of your customers? There are many different options that you can use when creating a segmentation. But let's look at the process that you can use to create a segmentation for your Lean Analytics project. The steps you need to use include:

- **==Define the purpose of your segmentation:==** You should first figure out how you want to use your segmentation. ==Do you want to use it to get more customers?== Do you want to use it to manage a portfolio for your current customers? Do you want to reduce waste, become more efficient, or something else?

Defining your purpose can make it easier to know how you should segment out your customers.

- **Identify the variables are the most critical:** These are going to help influence the purpose of your segmentation. Make sure that you list them out in order of their importance, and you can use options like a Decision tree or Clustering to help. For example, if you want to do a segmentation of products to find out which ones are the most profitable, you would have parameters that are revenue and cost.

- Once you have your variables, you will need to **identify the threshold and granularity of creating these segments.** These should have about two to three levels with each variable identified. But sometimes you will need to adapt based on the complexity of the problem you are trying to solve.

- **Assign customers to each of the cells.** You can then see if there is a fair distribution for them. If you don't see this, you can look for the reasons why, or you can tweak the thresholds to make it work. You can perform these steps again until you get a distribution that is fair.

- **Include this new segmentation in the analysis** and then take some time to look it over at the segment level.

Cohort Analysis

The Cohort Analysis is a test involves comparing sets of data using a time bucket. In this test, there will be differences in behavior between customers who arrived at the free trial stage of your process, versus those who showed up at the initial launch, and then those who are in the full payment stage.

Each of these is significant because it helps you to figure out which customers are likely to come back and be full-fledged customers when in the future. Those that show up in the initial stages when the product is free are often not the customers you are going to see when sales actually start. They may have just wanted to try it out and didn't really have an investment in the product.

Those that are in the later two stages can be customers who are better for you to work with. They will be the most interested in the product because they invested some money to get it. You really want to study these using the cohort analysis to figure out who your real customer base is and how they behave so that you can better market to them later on.

A/B Tests

A/B testing is a process where you examine an attribute between two choices. This could be something like an image, slogan, or color so that you can figure out which option is the most effective choice.

Let's say that you had two products that you are comparing and you want to find out which ones customers liked the best. Did they choose one product over the other and why? Did they respond better to the choice that was in green or the one in blue?

For this test to really work, you must assume that everything else is going to stay the same. So, it would have to be the same product, but there is one variable that is different between them. You could put up a website, for example, and have a red background on one version and a yellow background on another. Then you could use A/B testing to figure out which one the customer responded to the best out of those choices.

In addition, you can also work on multivariate analysis. This is pretty much the same thing, but instead of going through and testing out one attribute, you will go through and compare several changes against another group of changes to see which is the most effective. This one will require there to be a few changes in the second product compared to the first to be the most effective.

There are several keys that you need to have in place when you are ready to do an A/B test. These include:

- Know the reason that you are running this A/B test.
- The item that you are testing needs to be noticeable to the audience. If you make a minor change that no one is going to notice, then your results are not going to be that reliable.
- Stick with testing just one variable at a time. If you go through and do multivariate testing, or test more than one thing at a time, you will run into trouble. You may not know for sure which variable is causing the changes you see.
- Your test needs to end up being statistically significant. This means that it must have a sample size that is big enough to test and know that the results are valid within a certain margin of error.

Let's take a look at an example of how to do this. We are going to use this test on a website that you are trying to improve. There are two main ways that you can do this including:

- You will test the pages on separate pages.
- You will use JavaScript to conduct the test inside the page, so you don't need two different URLs to do it.

The first option is going to mean that you will need to have two different URLs for the pages that you are testing out. You can make them similar names, but make sure that there is some way that you can keep track of them and not get the two confused.

With the second option, you will need to have some experience working with JavaScript. You can then place some of this code on the website so that it can dynamically serve one option or the other.

The method that you choose is often going to depend on the one that you like the most and which tools you want to use. Both of these will give you some valid results, but you will find that implementing each of them takes a different amount of time to set it up.

Chapter 7: Step 1 of the Lean Analytical Process: Understanding Your Project Type

Now that we have taken a look at some of the different part of Lean Analytics, it is time to take a closer look at how the process works. These can help you to get started with the Lean Analytics stage for your business and ensure that you are getting the most out of Lean.

The first step that we are going to look at is ==understanding your business or your project type.== How are you supposed to pick out the right metrics if you have no idea what kind of business or project type you are working on? You must really understand the project at hand so that you can choose a fantastic metric that can show you results.

There are ==six general business types== that you can fit into, and they all will have metrics that are going to work best or matter the most, for each one. If you see that your business or project is on this list, your job will be simple. You just need to focus your attention towards understanding the priorities of what needs to be measured. This can include in-depth external research.

However, if you have a business that is not on this list, this doesn't mean you are out of luck and can't do anything. You can just use some of the information that is in this chapter as an example and build up your own understanding and metrics from this chapter.

E-commerce

The first type of business is going to be e-commerce. These are growing like crazy right now as many customers are looking for the things they want to buy online more and more. And many companies find that they can make large profits by offering their products and services online to these customers. An e-commerce business is going to be any that has their customers buy from a web-based store. This could include businesses such as Expedia.com, Walmart.com, and more.

The strategy for this type of business is that you need to understand the customer relationship that you want. This means that you are going to focus either on new customer acquisition or customer loyalty? You have to decide between these two because this is going to help with all other decisions that you make with this type of business.

There are many metrics that you can choose to go with in an e-commerce business. Some of the typical ones that other companies have chosen in this industry include:

- Inventory availability
- Shipping time
- Mailing list and how effective it is
- Virality
- Search effectiveness
- Shopping cart abandonment
- Revenue that you make on each customer
- The amount you spend to get new customers
- Shopping cart size
- Repeat purchase
- Conversion rate

The best metrics

Of course, there are several metrics that will work the best and will provide you with the best return on investment, when working with an e-commerce site. The best metrics to use here include:

- **Conversion rate:** This is the percent of all visitors to your site who also purchase something. The average conversion when it comes to online retail is 2%. There

45

are some that can do better though. For example, Tickets.com is over 11%, and Amazon.com is at almost 10%.

- **Shopping cart abandonment:** It is typical that 65% of the shoppers to your website are going to abandon their carts. Many of these are because of the high costs of shipping, and others are from the high price of all the items in their cart. You should definitely take some time to analyze any shopping cart abandonment that is happening in your business so that you can learn why you are losing these customers.
- **Search effectiveness:** The majority of your buyers are going to have to search to find what they need. If you make your search more effective, it can help your customers find what they want, rather than having them leave in frustration. Remember that about 79% of your total shoppers will use the search engine for half of the goods they want.

Software as a service

These types of companies are going to sell software in downloadable form or as a subscription. This can be things such as Skype, Evernote, Basecamp, Adobe, and more. They are not selling a physical product to someone, but these software programs are still pretty important for most people to get work done or to do other things on their computers.

==The strategy with this one is that most software is going to consist of products that are on a subscription which means that retaining the customers is important.== Your success is going to really depend on building up a loyal base of customers faster than those customers disappear.

There are some metrics that you can use to make this happen. Some of the ==most common metrics== that are used with this type of business model include:

- Reliability and uptime
- Upselling
- Virality
- ==Customer churn==
- ==Customer lifetime value==
- Cost of getting new customers
- The amount of profit you make per customer

- User conversion
- User stickiness
- User enrollment
- User attention

The best metrics

Just like last time, you are able to use any of the metrics that are above, but there are some that could be the best for helping you reach your overall goals. Some of the best metrics to use with a software company includes:

- **Paid vs. free enrollment:** You will find that your enrollment rate is going to change depending on whether or not you asked for credit card information in the free stage or not. The former is going to get an average signup rate of 2 percent, and then 50 percent often end up buying. When you do not ask, the average may increase to ten percent, but only 25 percent purchase the product.
- **Growing revenues and upselling:** Some of the best software providers are able to get 2 percent of their paying subscribers to increase what they pay each month. Being able to grow your customer revenue by 20 percent in a year can be achieved if you work towards it.

- **Churn or attrition rate:** This is the percent of your customers who are leaving. Going across the industry, the top companies usually have an attrition rate between 1.5 and 3 percent each month. If you have a percentage that is higher, then you need to find ways to make the customers stay.

Mobile app companies

These are companies that are going to provide apps to be used on mobile devices like Android and iPhone. Some of the companies that can fall under this category would be ones like WhatsApp and Instagram.

The strategy that you want to go with here is to find the right target audience. There are a lot of ways for your app to make money, but you will find that the majority of your revenue is going to come from a smaller group of customers, rather than from the population as a whole. You should focus your analysis as well as the metrics you use to help you better understand those customers.

There are many metrics that you are able to use as an app company. Some of the most common options include:

- Customer lifetime value
- Churn rate
- Ratings click-through
- Virality
- The revenue you make from each paying user
- The revenue you make for each user
- Percentage of users who end up paying
- How much it costs to get the customers
- Launch rate
- Downloads

Best metrics

Of course, there are many metrics that you can choose to look at when it comes to being an app company, but a few of them are going to provide you with the most information and can help your business to really grow. Some of the best metrics you can use include:

- **Downloads and the app launches:** The number of people who download the product and then activate it will fit in here. It is known that quite a few people who decide to download an app will then never activate it or use it at all, especially if the app is free.

- **The cost to get new customers:** You can follow a general rule to have a budget of 75 cents per user in your marketing initiatives to help attract new customers. You should always make sure that the cost to get new customers is lower than what you will earn on them. So, if you will only earn 50 cents on a customer, then you shouldn't spend 75 cents on each one.
- **The average revenue you earn per customer:** This is often going to be determined through the business model. For example, Freemium apps, or apps that you receive revenue from engagement in the app, will often have a higher revenue per user compared to those that are premium apps.

Media site companies

If you are in this industry, you have a website that is going to provide some information, such as articles, in return for earning advertising or any other type of revenue. These would include most blogs and other sites like CNN.com, CNET, and more.

Media sites need to really understand the source of their revenue. It is not coming directly from their readers or the people who use their "product," but it is coming from advertisers who are trying to reach those readers. So, if you are a media site company, you would get revenue from affiliates, click-based advertising, display advertising, and sponsorship. You would want to design your key metrics to work for this.

Some of the different metrics that you can choose to work with for a media site company include:

- Page inventory
- Pages per visit
- New visitors
- Unique visitors
- Content and advertising balance (you don't want too much advertising on the page, or it takes away from the content and keeps the customer away).
- Click through rates
- Ad rates
- Ad inventory
- Audience and churn

Best metrics

- **Click through rate:** This is the number of users that are going to click on a link out of all the users who check out the page. The average click-through rate for a paid

52

search in 2010 is 2 percent, but some companies can get higher. If you see that you are at one percent, then it is time to make some changes. But if you are above that number, you are doing really well.

- **Engaged time:** This is how long your reader will stay on the site and look through the content and the ads. Most media sites are going to aim for 90 seconds for content pages, and a little less with landing pages. If you find that your visitor is not spending more than a minute on the content pages, then it is likely your content is not engaging them.

- **Content optimization for media:** This one means taking the content that you already have and changing it so that it works on other venues, such as podcasts and video. You should track how others are using the materials you have because this can help you find some new opportunities to use.

User-generated content business

If you have a community that is engaged, they are going to contribute free content. And this same engagement is going to provide you with ads as well as other revenue sources. Some examples of companies that work with this include forums, Wikipedia.com, Reddit.com, Facebook.com, and Yelp.com.

The strategy that you should use is one that takes into account user engagement. This business is going to be successful when its visitors become regular contributes, and they interact with others in the community and provide quality content. User engagement ~~tiers~~ *tries* to measure involvement can be good as well.

Some of the different metrics that you may want to use with user-generated content include:

- Notification and mail effectiveness
- Content sharing
- Value of the content that is created
- Content creation
- Engagement funnel changes
- How many engaged visitors you have

The best metrics

- **Time on the site each day**: Here you are going to measure how long the average user is on your site and engaged on a typical day. This is a good thing to measure for engagement and stickiness. The average number is about 17 minutes a day, though Facebook is usually an hour, and Tumblr and Reddit are 21 and 17 minutes respectively.

- **Spam/Bad Content:** With these kinds of communities, you need to make sure that good content is always uploaded. You will have to spend time and money to keep bad content and fraudulent content off the site. You can measure what you think is good and bad and then build up a system to help keep up with this. You can also spend your time watching out for quality decline and then fix it before it ends up ruining your community.

Two-sided marketplace business

These kinds of businesses are going to connect buyers and sellers, and they will earn a commission on the work. It is kind of a variation of the e-commerce store. Some options of this would include Priceline.com, Airbnb.com, Ebay.com, and Etsy.com.

The strategy with this business is that you need to be able to attract in two different customers, the buyers and the sellers. The best bet is to focus on those that have the money to spend first. If you can find a group of people who want to spend their money, then those who want to make money will pretty much line up to do it.

Some of the metrics that you are able to use when it comes to a two-sided marketplace business include:

- The volume of sales and the revenue you earn
- Pricing metrics
- Ratings and any signs of fraud showing up
- Conversion Funnels
- Search effectiveness
- Inventory growth
- Buyer and seller growth

Best metrics

- **Transaction ~~side~~** [size]**:** Sellers usually won't have the money or time to analyze pricing and the effectiveness of their copy and pictures. As the owner, you will have the aggregate data from all your sellers, and you can use this information to help them with this analysis. Transaction size is the same as the purchase size, and of course, it is going to differ based on your business type. You should help your sellers measure it so that they can understand the behavior of your buyer and use it to sell more items.
- **Top 10 lists:** You can make top ten lists to help your buyers find the best products, and your sellers to know

==what is going on in the industry and what they can do to be more profitable.==

As you can see, there are many different types of businesses out there. And it is likely that your business is going to fit somewhere in this list. If it does, then there is an outline that you can use for developing a good strategy. Even if you don't, you can combine a few of these strategies to help you come up with the metrics, and the plan, that you need to succeed.

Chapter 8: Step 2: Determine Your Current State

Now that you know which business type you are in, it is time to move on to the second step of Lean Analytics. This one is going to require you to determine which innovation stage you are in right now. ==The one metric that means the most to you right now is a function of time.== It is going to change as your project keeps moving on through the different stages of innovation. There are several different stages of innovation that you can work with. These include:

==Stage 1:== *Empathy or is this a real problem?*

In this stage, you are going to ==identify a problem in your business and then get inside the head of your potential user.== You should be in their shoes and understand why there is a problem and what they are thinking. You may need to spend some time talking to potential customers to help with this stage. The more that you are able to talk to your potential customers and others in the market about the product or service you want to offer, the better off you will be. This can give you some real insights that can drive your business forward.

You need to focus on any metrics that are going to help you to determine whether or not the problem is harmful to your business. The metric needs to also determine if there are enough people who care about this problem. If only a few people see it as a problem, then it probably isn't worth your time taking care of it. But if a big percentage sees this as a problem, then it is something to take care of. You can also use metrics that will see what the success rates of your existing solutions are and if you need to change some of them.

Stage 2: Stickiness or do I have a good solution?

In this stage, you are going to start by making a Lean prototype of your solution to the problem you found in the previous step. You have to ask yourself whether or not people will pay for this. This is when you can gain feedback from small focus groups and testers. Based on that information, you can make adjustments and changes to the solution until you get it right.

You are going to need to focus on any metric that proves your solution will encourage the user to engage and also come back to your business.

Stage 3: *Virality or does this solution provide value to enough people?*

Once you have a solution and a product and they are seen as effective, you need to decide whether its value adds enough that the customer will tell others about it. Remember that word of mouth endorsements are valuable as a precursor to growth measurement and as free advertisement for the business. You can work for endorsements that are either natural (the customer enjoys the product or service enough that they just give out recommendations to their friends) or ones that are incentive-based (such as giving the product for free or at a discount).

For this one, you need to focus on metrics that are able to measure out if you are getting any new customers from your existing ones. And you want to know how many of these referrals are happening. You can also take a look at metrics that can check for how long it takes for news to spread or the cycle time.

Stage 4: Revenue or can I make this profitable?

Now you need to work on how much revenue you can expect from selling the product or service that you created in the last step. You can work on prices, standardization, control costs, and margins. You need to take the time in this step to prove that you are able to make money in a self-sustaining and scalable way, or this is not the solution for you.

This one is going to need you to focus on metrics that can tell you the net revenue that you are able to earn for each customer. The net is going to be the revenue that you make per customer minus the amount you spent to get that customer.

Stage 5: Scale or can we expand to a bigger audience?

Now that you have a product, you showed that it is effective, and you have a business model in place to show that it is going to be functional and profitable, you can now invest and expand it into new markets. This can include new geographies, channels, and audiences as well.

If you are dealing with a project or business that is oriented on efficiency, you need to focus on metrics that are able to reduce costs. If you are working in a business or project that is differentiation oriented, you will want to focus on metrics that will track margins for you.

What can I do with these innovation stages?

Now that you know a little bit more about these innovation stages, it is time to figure out where you are and learn what you can do with each one. The steps that you should take from here include:

- Look through the stages above and determine where your business or project is right now.
- Refocus on the things that you should be measuring at the stage you are in.
- If you find that your project does not fit into this framework, then it is important to remember that all innovative endeavors are going to follow a pattern of stages as well as maturity through time. Are you able to borrow this framework and leverage it in some manner so that you can figure out what stage your project is in right now? This can really make a difference in helping you to understand what you really need to be focusing on right now.

Knowing where you are in the innovative stage can make a big difference. When you look at the five stages above, you will have a clear outline of what you need to focus on and what needs to be done to keep you moving forward. If you have no idea where you are right now, then how are you supposed to know what steps to take to get to the next level? Always have a good idea of where you are in the innovation process, and then you have a clear picture of where you should go next and can keep on track.

Chapter 9: Step 3: Pinpoint the Most Pressing Metric

To help you to be successful with any type of innovative project, the key is to focus. Consequently, you can't spend your time on too many metrics because this is going to make you feel distracted and you are going to lose all your focus.

In the first few stages of innovation, it is often best to reduce the number of metrics that you track. If you can, you should focus on just one metric, the one that matters the most right at the moment. The metric that is the highest-priority is usually related to the most important project or business need.

For example, a subscription software company may be in the virality stage, and they are trying to gain traction with it. They may decide that the net adds metric is the one that will help them out the most. Remember that Net Adds = Total of New Paid Subscribers − Total That Cancelled.

There may be other metrics that your company can use, but you need to just focus on one. You will need to figure out what problem is the most pressing or the most important right now, and then go with a metric that fits with this the most.

How can I find that one metric?

Some of the steps that you can use to find the one metric that matters the most right now include:

- Write down the top three to five metrics that you really like, and you often track.
- How many of these metrics are actually any good and help you out?
- How many do you use to make business decisions? How many of those would actually be vanity or false metrics?
- What stage are you at with the business or the project? Do you really understand what matters in the business model? Can you discard any of the metrics that aren't really adding value to you right now?
- Are there any other metrics that are not on your list that you can think of and that you think could be more useful right now?
- Once you have written down all those metrics, you can go through the list and cross off any bad or false metrics. Add any new good ones that you think of on the bottom.
- Now that you have a list, you should go through and pick the one metric that you absolutely can't live without to help you with the project in its current stage.

What to do after optimizing that one metric

Once you have the metric and the project at a level where you are happy with the numbers, you must remember that you will need to continue measuring them. You never know when that project or that metric will need to be changed up again to help you in the future. But you can rest assured knowing that the process is now controlled and optimized. What this means is that you are now at a point where you are achieving a certain level of results.

Now you are able to go back to your list and work on the metric that is the next highest priority. This is going to be the next highest priority of your business, or the next biggest project need. You can review through the innovation stage you are currently at and the business or project type, and then you can determine which point of interest you should focus on.

Remember that the goal here is to only work on one metric at a time. There may be several metrics that need to be addressed in your business. Keep your focus on one at a time.

Sure, you can go through and write out a list of the different things that need to be addressed at some point, and Lean Analytics is a good time for this because it can help you see what problems are there. But you should pick the one that is the most pressing and work on that one first.

After you have time to complete the Lean Analytics stage on this problem and you have a winning strategy in place for it, then you can move on to the next step of picking out a new project to work on. You can implement this process on as many projects as you would like. Just make sure that you are only working on one at a time.

Chapter 10: Tips to Make Lean Analytics More Successful for You

Getting started with Lean Analytics is something that can take some time to get used to. It is going to provide you with great results and a winning strategy that is sure to get you ahead. But for those who are just starting out with this stage, or who are just getting started with the whole idea of the Lean system, you may need some help to get going on the right foot. Here are some great tips that you can follow to ensure that you are doing well with Lean Analytics and to ensure it is as successful as possible for you:

- **If you are doing an A/B test, you need a lot of users:** You are not going to get any good results from your A/B test if you don't have a lot of users to help you out. This means that it is not going to work all that well if you are a small startup or if there are not a lot of people you can measure. You should have a minimum of 10,000 events before you attempt this kind of test. These events can include visits or people who use a feature. Make sure that you are able to get this many users to help you out before you get started.

- **Make big changes:** If you are not able to see the changes from a few feet away, it is likely that the people you are testing with the A/B test won't either. For example, A/B tested 41 different shades of blue. The results were not the best because there were just too many different shades and for most people, they looked too similar. You need to make big changes before you do an A/B test, or it won't work well for you.

- **Measure the tests properly:** You are not going to get the right results if you are not properly measuring the tests. You need to have the right metric in place. Also, make sure that you never stop a running test too early, or you may miss out on some of the important results that you need.

- **Use the tools that you need:** Lean Analytics has a ton of tools that you can use to make it successful. Make sure that you are properly trained to handle each part and that you don't miss out on some important tools that can make this more successful.

- **Know where your business is now:** How are you supposed to have any idea of what kind of project to work on and what metrics to use if you don't have a good understanding of your business? Make sure that you know the overall goals and vision of the business. This can help you to spot some of the problems that you need to fix and can make it easier to ensure that whatever changes you do decide to make are going to go along with what your business is all about.

- **Understand the different metrics**: You should spend some time looking at the different metrics that are available for you to use on your project. Each one can be great, but it does depend on the type of business that you are running and the project that you want to work with. You need to learn which metric is going to be the right one for you.

- **Add this into the Lean Support System**: Many people are fond of the Lean Support System. This allows them to get rid of a lot of the waste that their company may have, and can make them more efficient. But you need to do Lean Analytics first to see success. This helps you to gather the information that is needed and then sort through it and analyze it. Then you can use this information to come up with the best plan to handle your problem. If you just jump into a strategy without the resource, it is likely that you won't see results at all.

- **Focus on the main problem first**: If you are like many businesses, there are probably many problems that you need to solve. But you don't have the time and resources to do all of them at the same time. When you get started with Lean Analytics, you must figure out what the main issue is, the one that will have the biggest impact on your profits, and work with that first. Once you have successfully implemented Lean Analytics and worked on the problem, then you can go back and see if there are any other problems that need to be addressed.

- **Get rid of the waste:** Remember that the most important thing that you will do with the Lean system is get rid of waste. And the data that you collect in the Lean Analytics stage is meant to help you to find the waste and learn how to get rid of it. Take a look at some of the most common types of waste that businesses may experience (and that we listed in an earlier chapter) to give you a good idea of where to start.

Lean Analytics can be a great way for you to get a strategy together that will help your business become more successful. If you follow these tips and some of the strategies that we talk about in this guidebook, you are sure to see some amazing results in no time.

Conclusion

Thank you for making it through to the end of *Lean Analytics: The Complete Guide to Using Data to Track, Optimize and Build a Better and Faster Startup Business*, let's hope it was informative and able to provide you with all of the tools you need to achieve your goals.

The next step is to start the process of implementing Lean Analytics into your own business. Learning how to make changes so that you can be more cost effective and provide better service to your customers all starts with Lean Analytics. This stage asks you to search for the data you need and analyze it so you know what step to take next. You can't come up with a plan for improving your business without the help of Lean Analytics to make it possible.

Finally, if you found this book useful in any way, a review on Amazon is always appreciated!